Foreword

Dedicated to all children, learn and have fun...!

I couldn't figure out why learning the times tables needed to be quite so difficult. Everyone said the best way was to just know the numbers, but I just knew there had to be an easier way to learn and teach them. I started researching how memory and recall worked, and I discovered how the mind uses color, imagination and association to recall intricate facts – fast! I decided to use these to teach my children the times tables, and it worked like magic! I've been asked to share the times tables stories and am sure it can work for your children too. They're a wonderful way to sit down and have fun letting both your imaginations go wild!

C. Montgomery

Instagram/TimesTablesMagic

Meet the Numbers

The stories are about number characters making up the times table. They lead onto a colourful story using imagination and exaggeration to give the answer, helping recall the answer quickly and easily. The story of the two butterflies, which is 3 x 3 will help you recall the balloon – the number 9.

During the stories, different examples are used to make the characters stand out, but you'll soon learn woodpecker without it's tree is still the number 12! The order of the numbers is important, so if it's 11 x 12, paint brush comes first , then butterfly and finally swan giving 132, enabling recall of the correct answer. All the tables up to 12 are in the book with the exceptions of 1, 2, 10, and 11 up till 10, but 11x11, and 11x12 are included. This is because these are easy recall tables and we wanted to keep it as simple for children as possible. Now to meet the characters....

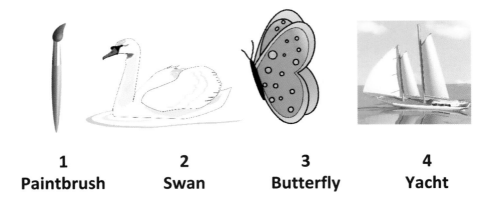

1	2	3	4
Paintbrush	**Swan**	**Butterfly**	**Yacht**

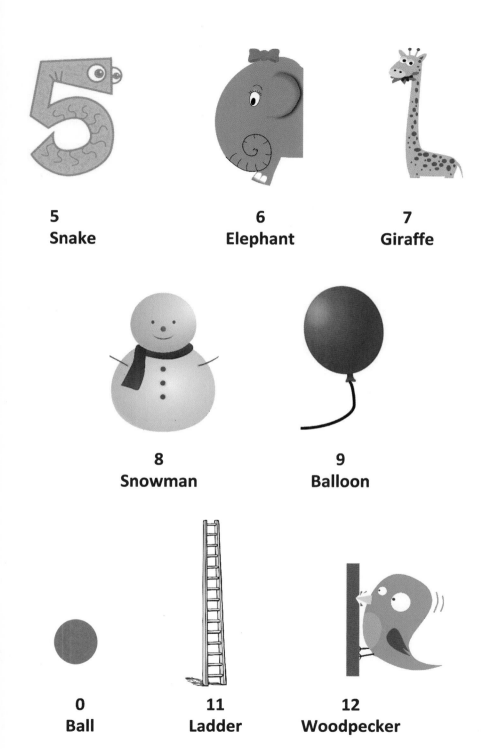

5
Snake

6
Elephant

7
Giraffe

8
Snowman

9
Balloon

0
Ball

11
Ladder

12
Woodpecker

3 x 3 = 9

Butterfly and the Butterfly

Butterfly and butterfly are pretending to swim butterfly stroke – there are crowds around them cheering them on! Suddenly there is a giant red balloon with a long tail in the shape of a 9 cheering them on! Balloon is 9

4 x 3 = 12
Yacht and the Butterfly

The yacht and the butterfly are having a race, yacht is panting *'you might be prettier but I'm much faster!'* He yells! But suddenly butterfly calls on her mates, she summons woodpecker from across the woods to jabber down the mast holding yacht's sails!

Woodpecker is the number **12**

5 x 3 = 15
Snake and the Butterfly

The snake is trying to snap at butterfly, who is dancing around his long wispy tongue. She sprinkles some magic dust on her antennae and they turn into a paintbrush to paint in the air - *'here's another snake you can play with of course!'* Paintbrush and Snake

are the number 15

6 x 3 = 18

Elephant and the Butterfly

Elephant is sick he's growing butterfly ears! He goes to the Doctor '*I'm feeling sick*' he says looking to the floor. There's a huge paint brush on the desk, Dr Snowman turns around, looking quite impressed. '*Worry not elephant a spoonful of this will fix you right up!*'.

Paintbrush and Snowman are the number **18**

7 x 3 = 21
Giraffe and the Butterfly

Giraffe has grown butterfly shaped spots! Whatever has happened to him! Giraffe looks down at the water and tries to wash himself off, *' swan can you help me scrub off?!'* Swan replies *'I can't so easily, let me see if I can draw on some new spots'.* So she sets to work with her paintbrush! Swan and

Paintbrush are number 21

8 x 3 = 24

Snowman and the Butterfly

Butterfly has landed on snowman's long carrot nose thinking how long it grows! Suddenly the sun appears and snowman melts down. Butterflies wings separate and transform into a swan! The antennae, as if by magic are drawn into a shape of a yacht! The picture has changed quite somewhat! The swan and the

Yacht are the number 24

9 x 3 = 27

Balloon and the Butterfly

Butterfly is pushing balloon along the water, swan wants to join in. But butterfly says *'you need a partner to swim in this race, I have balloon who do you have?.'* So swan calls giraffe who is nibbling leaves nearby, and giraffe swims in and swan jumps

on his back! Swan and Giraffe are the number 27

12 x 3 = 36

Woodpecker and the Butterfly

The Woodpecker and butterfly are having a discussion, woodpecker says *'do you think this is truth or fiction?'* *'I have seen butterfly ears, great big flappy ones, on an elephant!'*. Truth or fiction!?

Butterfly and Elephant are the number 36

4 x 4 = 16
Yacht and the Yacht

The yachts are having a race, both are whizzing along the water! But they're not the same shape.

One has a long straight paintbrush shaped sail, the other has a curved one rolling up like an elephant

trunk! Pencil and Elephant are the number **16**

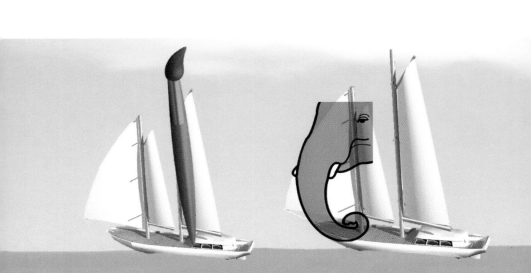

5 x 4 = 20

Snake and the Yacht

Snake is wrapped all around the mast of the sail, oh no yacht can't be racing if the snake won't leave! Everyone is telling snake to come off but he won't listen. At last an enormous swan comes flapping its soaring wings and throws a ball on snake so he can come off the mast and down to the ocean! Swan

and Ball makes the number 20

6 x 4 = 24

Elephant and the Yacht

The yacht is carrying elephant to the sea shore breezing along in the wind, but yacht is getting tired, elephant is heavy. So yacht asks a nearby swan to help push him along.

The swan hails down another yacht with her long wings and so the swan and the yacht push them to

safety. Swan and Yacht are the number 24

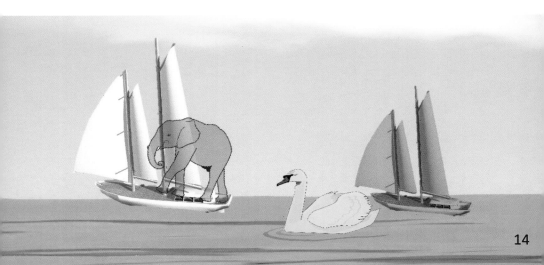

7 x 4 = 28

Giraffe and the Yacht

Yacht is carrying giraffe to the shoreline, but giraffe has forgotten his spots! He asks the nearby swans for help to fetch them.

A swan speeds off and returns with snowman on her back, he is carrying the spots, *'I've found them!'* he yells! Swan and Snowman are the number 28

8 x 4 = 32

Snowman and the Yacht

Yacht is carrying snowman safely to the shore, snowman is melting in the sun!

Butterfly sips some of the water, she is so thirsty. But then hops onto a nearby swan to help push the yacht along more quickly! Butterfly and Swan are

the number 32

9 x 4 = 36
Balloon and the Yacht

The balloon is flying high on the mast of the yacht, prancing around in the wind looking like a colorful spot!

They're racing fast, the wind in the sail, but look closely at the balloon – it has a picture of butterfly ears on an elephant! Butterfly Elephant ears is the

number 36

12 x 4 = 48

Woodpecker and the Yacht

The woodpecker is jabbering away at the mast of the yacht instead of its tree! Yacht needs to be rescued!

Another yacht is coming to the rescue being captained by snowman, pointing his carrot nose towards the yacht to be rescued. Yacht and

Snowman are the number 48

5 x 5 = 25
Snake and the Snake

Snake is dancing with another snake, they're doing a prance, it's so strange. They look like the shape of a swan, and then they change shape! Now they look

like just a single snake! Swan and Snake make 25

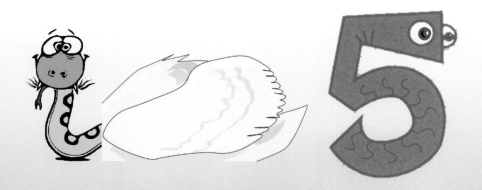

6 x 5 = 30
Elephant and the Snake

Elephant has lost his nose, instead, he's swinging snake around pretending he's his trunk!

Poor snake is bruised, there's an enormous one the shape of a butterfly on his head, it looks like it's juggling a ball instead of flying how strange. Butterfly and Ball are 30

7 x 5 = 35

Giraffe and the Snake

Giraffe and snake are measuring the length of giraffe's neck. Snake straightens himself out and lays across giraffes neck. Meanwhile butterfly is trying to measure snake's length! She flutters forward and sits herself down next to snake. But butterfly made a mistake for it's not safe! Snake calls his friend and they're chasing butterfly! Butterfly and Snake make 35

8 x 5 = 40

Snowman and the Snake

Snowman has been given a walking stick in the shape of a snake! Every time anyone walks past him he uses the snake stick to tip his hat.

His hat is shaped like a black yacht on his head, with a red ball on the top to really look great! Yacht

and Ball are the number 40

9 x 5 = 45

Balloon and the Snake

Snake is wrapped all around balloon! The string is secured to the ground with a weight. It looks like the shape of a yacht anchoring the balloon down! Snake slithers down and folds himself into the yacht!

Yacht and the Snake are 45

12 x 5 = 60

Woodpecker and the Snake

Snake is wrapped around the tree that woodpecker is jabbering away at. Suddenly woodpecker hears a loud noise and heavy footsteps. It must be elephant, there's a circus coming, and elephant is running along juggling

a ball! Elephant and Ball are the number 60

6 x 6 = 36

Elephant and the Elephant

Twin elephants are merrily playing skipping, but when you look closely their ears are the shapes of butterflies flapping in the wind.

Butterfly ears on an Elephant is the number 36

7 x 6 = 42

Giraffe and the Elephant

Giraffe and elephant are feeling sorry for themselves, one is tall and one is wide. *'Oh I wish I could be taller'* says elephant to giraffe and starts crying big drops of tears. Soon, there's lots of water and a yacht is sailing past, swan is onboard and gets ready to brush away the tears with her feathers! *'Be proud of who you are!'* she says.

Yacht and Swan are 42

8 x 6 = 48

Snowman and the Elephant

Snowman and elephant are making snow angels in the snow, but they didn't realise it was thin ice. Elephant falls through into cold water so snowman has to call the rescue yacht with captain snowman to rescue elephant! Whew!

Yacht and Snowman are 48

9 x 6 = 54

Balloon and the Elephant

Elephant is presenting a balloon to a young girl *'well done!'* he says. But she quickly starts yelling! Naughty snake has curled himself into the balloon and let his tail out! The girl's brother asks if he can swap the snake for his toy yacht and she happily agrees. Snake and

Yacht are 54

12 x 6 = 72

Woodpecker and the Elephant

The woodpecker is jabbering at a tree, but it starts yelling 'OW!' *'Oh no'* says woodpecker *'sorry, I thought you were a tree'*. *'Well I'm not, that's my elephant trunk you're jabbering!'*. Woodpecker looks scared and turns to jabber a tree. It's tall and yellow! 'OW OW!' yells the giraffe, *'that's my neck!'*. Oh heavens whatever next! Swan is looking on in dismay and uses her feathers to brush the pain.

Giraffe and Swan are **72**

7 x 7 = 49

Giraffe and the Giraffe

Giraffe and giraffe are playing twister, both of them want to win. They look at the awards table, there's a large beautiful yacht and a beautiful red balloon, oh how they wish they could win their lots! Yacht and

Balloon are 49

8 x 7 = 56

Snowman and the Giraffe

Giraffe has spots shaped like a snowman.
Snake is laughing in his hissing way at this.
Giraffe asks:

what shapes do you have on your body?'

'My scales look like elephant trunks!' snake

answers! Snake and Elephant are 56

9 x 7 = 63

Balloon and the Giraffe

Giraffe is being carried through the air by a big hot air balloon. He can't believe his eyes and is quite frightened! But the captain yells down at him, *'don't worry ole giraffe, be grateful you're not an elephant otherwise you could never see this view', and look over there at the stunning butterfly, so no more tears!'*

Elephant and Butterfly are 63

12 x 7 = 84

Woodpecker and the Giraffe

Woodpecker is jabbering away at giraffes neck much to his dismay. It is a very hot day, and woodpecker gets thirsty. He pulls out a snowman shaped ice pop that he can enjoy. He nibbles away dreaming of cruising down the water in his yacht.

Snowman and Yacht are 84

8 x 8 = 64

Snowman and the Snowman

Snowman is holding hands with his wife, they have been handed their gorgeous gifts by their best friend elephant, it is a honeymoon on a yacht to sail off into the snowset!

Elephant and Yacht are 64

9 x 8 = 72

Balloon and the Snowman

Snowman is holding onto a balloon, bobbing around on his arm. Snowman suddenly hears a popping sound 'sorry!' Says giraffe, *I didn't see and was trying to eat it!'* Snowman sobs *'oh I loved that balloon!'* Giraffe tells his friend swan what happened feeling ashamed. Giraffe and Swan

are **72**

12 x 8 = 96

Woodpecker and the Snowman

Woodpecker is making the final touches to the snowman, she is crafting him herself instead of jabbering at her tree!

She wins first for her sculpture! Winning a hot air balloon ride to see the elephant safari! Balloon and

Elephant are 96

9 x 9 = 81

Balloon and the Balloon

The two balloons are bobbing in the air, they join together to form the shape of a snowman then they come apart, suddenly a sharp paintbrush is coming to pop them - quick! They'd better hide!

Snowman and Paintbrush are 81

12 x 9 = 108

Woodpecker and the Balloon

Woodpecker is bursting as many balloons as possible. She can see the **paintbrush** being used to write down her score! There is a **ball** being juggled around and around, look closely it is **snowman** he's keeping the score, but juggling to stay calm!

Paintbrush, Ball and Snowman are 108

11 x 11 = 121
Ladder and the Ladder

The firefighters have to use two ladders to rescue the woodpecker. She was using a **paintbrush** to jabber the tree, it got stuck in the bark along with her beak! Woodpecker and Paintbrush are 121

12 x 11 = 132

Woodpecker and the Ladder

Woodpecker is being rescued by a ladder as she went too high in the tree. The fireman below is writing her name down with a **paintbrush**, but keeps shooing away a **butterfly** getting too near! Suddenly a swan flies over and gently plucks butterfly away. Paintbrush, Butterfly and Swan are

132

12 x 12 = 144

Woodpecker and Woodpecker

Woodpecker and woodpecker are having a race, who can fly highest! They look over to the winning table where their scores are being written in bright red **paintbrush**, they could win a **yacht** each! Paintbrush and **two** yachts are 144

Printed in Great Britain
by Amazon